Smail Lamine

General ecology

Smail Lamine

General ecology

Structure and functioning of the Biosphere

ScienciaScripts

Imprint
Any brand names and product names mentioned in this book are subject to trademark, brand or patent protection and are trademarks or registered trademarks of their respective holders. The use of brand names, product names, common names, trade names, product descriptions etc. even without a particular marking in this work is in no way to be construed to mean that such names may be regarded as unrestricted in respect of trademark and brand protection legislation and could thus be used by anyone.

Cover image: www.ingimage.com

This book is a translation from the original published under ISBN 978-620-6-71676-1.

Publisher:
Sciencia Scripts
is a trademark of
Dodo Books Indian Ocean Ltd. and OmniScriptum S.R.L publishing group

120 High Road, East Finchley, London, N2 9ED, United Kingdom
Str. Armeneasca 28/1, office 1, Chisinau MD-2012, Republic of Moldova, Europe
Printed at: see last page
ISBN: 978-620-8-02957-9

Dr. Smail LAMINE

Foreword

This General Ecology course is intended for students in the Preparatory Classes, 2e year Natural and Life Sciences at the Ecole Supérieure d'Agronomie - Mostaganem. This educational document in no way replaces the books on the subject available in the library. It is simply a summary of a few choice titles, old and new, covering the curriculum of the teaching unit in question.

The course is presented in five chapters, each sufficiently illustrated to facilitate comprehension. The first chapter focuses on a few general points about ecology, namely the notion of an *ecological system* and the various fields of intervention. The second chapter presents *the environment and its interactions with the various elements that surround it*. The third chapter deals with *abiotic factors, i.e. the complexity of climatic and edaphic elements*. The fourth is devoted to *biotic ecological factors*, covering all existing interactions between different living beings. Finally, the fifth deals *with the structure and functioning of the various types of ecosystems* that characterize our biosphere.

I hope that this handout will help Biology students to understand and master the General Ecology module and thus successfully complete their 2e year Classes Préparatoires SNV. I also hope that this handout will provide Ecology teachers with a practical and effective working tool. I would be very grateful if readers would let me know of any possible errors or criticisms.

Dr. Smail LAMINE

Table of Contents

Preamble

This course handout for the General Ecology module is intended primarily for students in the second year - Classes Préparatoires (Semester 4 - First Cycle - SNV) at the Ecole Supérieure d'Agronomie - Mostaganem.

EU Methodology

Hours: 67h 30

Coefficient :

Credit :

Teaching objectives: General description of the science, structuring and functioning of different ecosystems. With this set of general theoretical notions taught, students will gain an in-depth idea of the place of living organisms in ecological systems; their action on the environment; and the action of the environment on their development and various adaptations.

Language of instruction: French / English

Type of course : Lecture

Assessment methods: 75% EMD exam and 25% continuous assessment (TD exam + field trip report).

Year: L2

Semester: S4

Recommended prior knowledge: Biology, Ecology and Zoology

Document contents :

Part 1: General ecology ;

Part 2: The environment and its elements ;

Part 3: Abiotic factors ;

Part 4: Biotic factors.

Part 5: Ecosystem structure and function

Specific objectives

On completion of this course, students will be able to :

■ Know and understand the specific terminology of the module in question;

■ Demonstrate the main mesological parameters of various terrestrial and aquatic ecosystems;

■ Know how to analyze the biodemographic processes involved in the dynamics of populations and settlements in different ecosystems;

■ Work with decision-makers to manage natural resources rationally and sustainably, and implement ecologically sound practices (**e.g.** sustainable development).

This document is based primarily on information gathered and synthesized from various references on fundamental and applied ecology.

To evaluate this course handout, we proposed Dr. BOUZID Khadidja of the Ecole Supérieure d'Agronomie de Mostaganem and Dr. GHARABI Dhia of the Université Ibn Khaldoun- Tiaret.

General introduction

In a field of research as vast as biology, there is almost always a dominant discipline, marking a given era, such as systematics in Linnaeus' time, physiology in the 1830s to 1850s, evolution and phylogeny in the 1860s and 1870s, genetics in the first two decades of the twentieth century... molecular biology since the 1950s and *perhaps today ecology*.

From the outset, ecology as a science has been built around the rather mad project of considering our surroundings as a "whole". In fact, the term ecology was coined by Ernst Haeckel in 1866, referring explicitly both to the relationships that organisms have with each other and to the relationships between organisms and the physical and chemical characteristics of their habitats.

During the second half of the nineteenth century and the first decades of the twentieth, a great deal of ecological research was undertaken, establishing a set of fundamental notions specific to this discipline. Thus, the concept of *biocenosis* (or *biocoenosis}* was developed by **Mobius as early as 1877**, and that of ecosystem by **Tansley in 1935**.

Despite the fact that the main issues in ecology were envisaged as early as the 1920s, over the last few decades there has been a definite evolution in the major concerns of the discipline.

Ecology covers a wide range of fields. Initially, it focused on individual species (oak ecology, fox ecology, etc.), where most of the research carried out concerned the analysis of the action of ecological factors on living beings *(factorial ecology* or *autoecology)*. Subsequently, research focused on more advanced levels of organization, such as populations *(demoecology}* or communities *(biocenotics* or *biocoenotics)*.

The study of the structure and functioning of ecosystems, in particular synecology, and of the biosphere as a whole, underwent fruitful development from the 1960s to the 1980s.

Finally, over the last three decades, the introduction of computer tools has led to the emergence of digital ecology, the aim of which is to model and simulate ecological systems.

Ecology is thus a mature science, producing theories - simplified representations of the world - that are valid in virtually all situations, and that lead to an ability to script the future or reconstruct the past. What's more, it is no longer just a discipline pursuing cognitive objectives; it has become a socially-involved science. While it is built on its own concepts and methods, it is also a crossroads science, drawing on many other disciplines, both living and non-living.

Chapter 1: General ecology

Introduction

Modern ecology is structured around two fundamental axes: the study of the dynamics and functioning of populations and populations, and the study of the dynamics and functioning of ecosystems and landscapes. The living world is structured according to levels of organization of increasing complexity (molecules, organelles, cells, tissues, organs, individuals, populations, communities).

1.1. Definition

Etymologically, the term ecology (from the Greek *oïkos*: habitat and *logos*: study) means the science of habitat. It was coined by the German biologist ERNEST HAECKEL in 1866 to designate the science that studies the existing relationships between living beings and their natural environment. According to DAJOZ (2006) & FAURIE *ET AL.* (2011), this term designates the global science whose object is the study of the conditions of existence of living beings and the interactions of all kinds between these living beings and their environment. It also highlights the relationships that living beings, including man, have with each other and with their environment.

As an applied science, ecology develops and implements the theoretical and practical knowledge on the basis of which most of the problems linked to safeguarding, managing or exploiting the biosphere's ecosystems and renewable resources should be posed and then resolved (BARBAULT, 2008).

1.2. Areas of intervention

Ecology occupies a special place in the biological sciences. It is a holistic (global) discipline. Historically, the development of this discipline actually began with the study of the action of ecological factors on isolated plants or animals. This area of the discipline is known *as **autoecology*** (or ecophysiology).

Later, research focused on populations *(demoecology)*. For many schools of ecological thought, particularly the Anglo-Saxon school, the minimum level of organization covered by ecology is the population. However, the most specific part of the latter corresponds to the higher levels of the pyramid (Figure, 1), namely the study of ecosystems *(synecology)*, which represents a major part of modern ecology, the aim of which is to study the structure and functioning of ecosystems.

On a larger spatiotemporal scale, we come across complex systems made up of several ecosystems that are adjacent to each other (i.e. neighboring or in close proximity to each other), known as landscapes. They constitute a higher-order entity that is the subject of specific developments under the term landscape ecology.

Moreover, the most complex level of biological organization studied by ecology is the biosphere and beyond, the ecosphere, the study of which is the subject of global ecology (FISCHESSER & DUPUIS-TATE, 2007; RAMADE, 2008 & SOTTIAUX, 2008).

Finally, the levels concerned by ecological studies are essentially: the individual, the population and the settlement.

Individual: This is a functional biological system which, in the simplest case, is reduced to a single cell. During its growth, the organism must adapt to the various conditions of the environment in which it lives.

Population: A group of individuals of the same species living in a given territory at a given time, characterized by a structure, organization, functioning and evolution, which are themselves controlled by the environment.

Settlement or community: is the set of populations in the same environment, animal settlement (zoocenosis) and plant settlement (phytocenosis) that live in the same mesological conditions and in close proximity to each other.

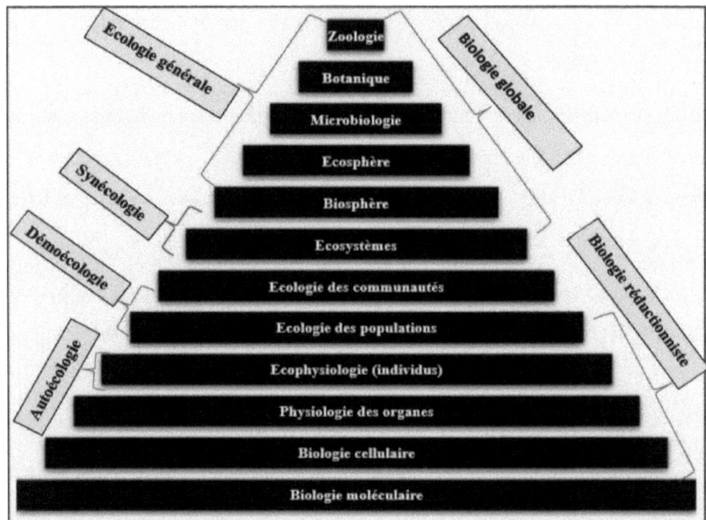

Figure 1: Schematic representation of the hierarchy of biological sciences in ascending order.

Figure 1 shows the hierarchy of the biological sciences. At the top of the pyramid are the traditional, classical subdivisions (Zoology, Botany, Microbiology, etc.). In the other parts of the pyramid, moving upwards from the base of the diagram, we see increasingly complex levels of organization of the living world, from the simplest level (individual) to the most complex (Biosphere), via the intermediate level (population) (RAMADE, 2008).

1.3. Notion of ecological systems: ecosystem

The object immediately given or accessible to the naturalist is an *individual*. Individuals, initially perceived as isolated in nature, only make sense to the ecologist through the *system of relationships* linking them to other individuals and to their physical and chemical environment (Figure 2).

The fundamental unit, the elementary component of ecological systems, depends on the objective of the study and the state of knowledge in the field. For example, we may be interested in the dynamics of a particular population, and so define the system to be studied by the network of direct or indirect relationships it has with the other biotic and abiotic components of its environment. In the schematic example shown in Figure 2, the latter have been grouped into a few fundamental compartments according to the type of interaction they have with the "central" population: predator, prey, competitors.

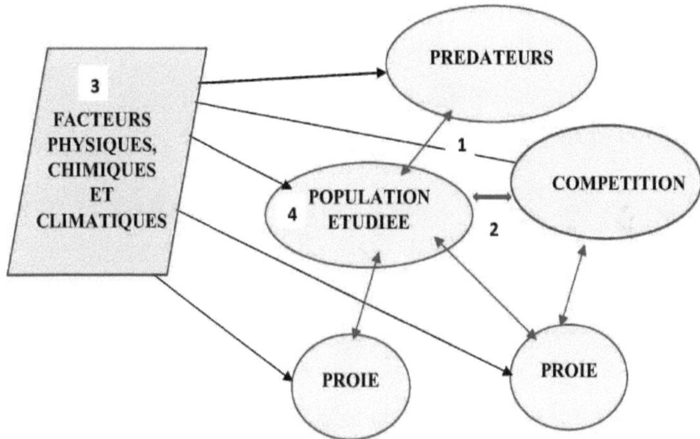

Figure 2: Schematic representation of an ecological system.

Natural populations are never isolated: they may interact with each other in various ways - predation **(1)**, competition **(2)**, cooperation (not shown) - and are subject to the physico-chemical factors of the environment **(3).**

1.3.1. The ecosystem concept

The pond is an example of an ecosystem. This notion is often considered as the central dogma of ecology, which can be compared to that of the genetic code in molecular biology. It is also defined as a biological system formed by two inseparable elements, the biocenosis and the biotope (Figure, 3) (EL ABOUDI, 2014).

A biocenosis: is a set of communities or living beings of all species, plant and animal, coexisting in a defined space called a **biotope.**

A biotope: is a set of elements characterizing a determined and uniform physico-chemical environment that is home to a specific flora and fauna (**the biocenosis**).

A species: a group of living beings that can reproduce with each other (**interfertility**) and whose descendants are fertile.

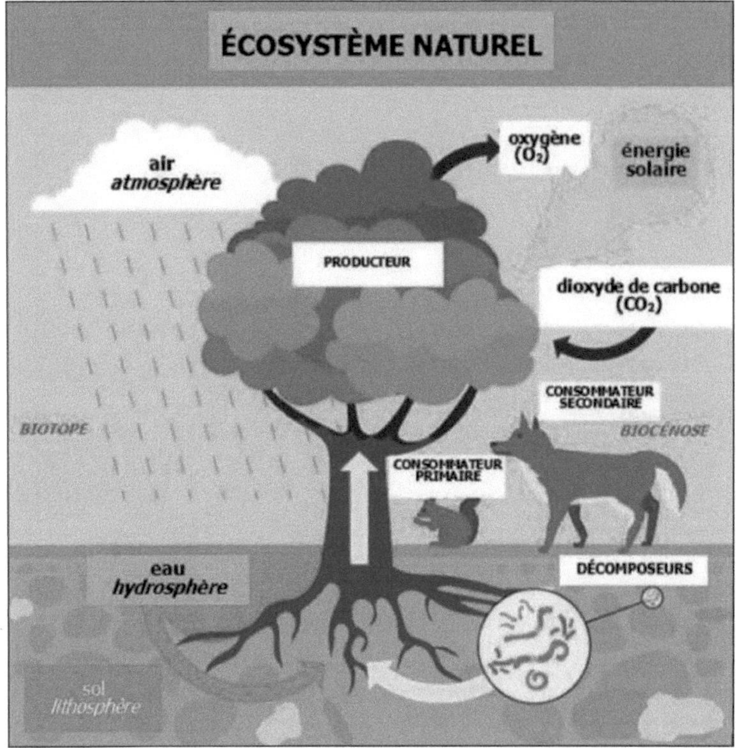

Figure 3: Example of a natural ecosystem

Biocenosis + biotope = an ecosystem

The biosphere: is the thin surface layer of the earth in which life has been able to flourish and sustain itself permanently (RAMADE, 2009). It partly covers three of the compartments that make up the earth: the atmosphere, the hydrosphere and the lithosphere (Figure 4). The biosphere is characterized by its low vertical development in relation to the earth's radius of 6,300 km, and by the opposition between the marine and terrestrial environments.

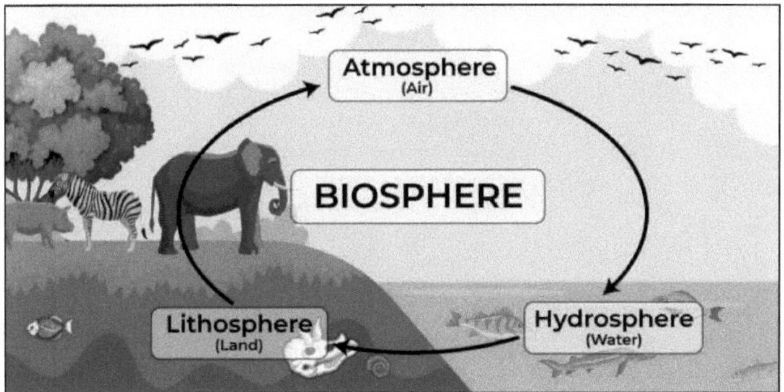

Figure 4: Schematic representation of biosphere compartments

1.3.2. Some examples of ecosystems

The notion of ecosystem is multiscalar (multi-scale), meaning that it can be applied to portions of the biosphere of varying dimensions: a lake, a meadow, or a dead tree...

Depending on the scale of the ecosystem, we have :

- a <u>micro-ecosystem</u>: e.g. a tree;
- a meso-ecosystem: e.g. a <u>forest</u> ;
- a macro-ecosystem: e.g. a region.

Ecosystems are often classified by reference to the biotopes concerned. We speak of :
- Continental (or terrestrial) ecosystems such as: forest ecosystems (forests), grassland ecosystems (meadows), <u>agro-ecosystems</u> (agricultural systems);

- Ecosystems of continental waters, for Ientic ecosystems of slow-renewing still waters (lakes, marshes, ponds) or <u>lotic ecosystems</u> of flowing waters (rivers, streams);
- Ocean ecosystems (seas, oceans).

Ecosystems are classified according to their size: micro-ecosystem (dead tree trunk, small island...), meso-ecosystem (forest, pond...) and macro-ecosystem (ocean, desert...).

Biomes are defined as homogeneous biogeographical groupings of ecosystems by climatic region covering a vast area (tundra, taiga, steppes, deserts, etc.) (Figure, 5).

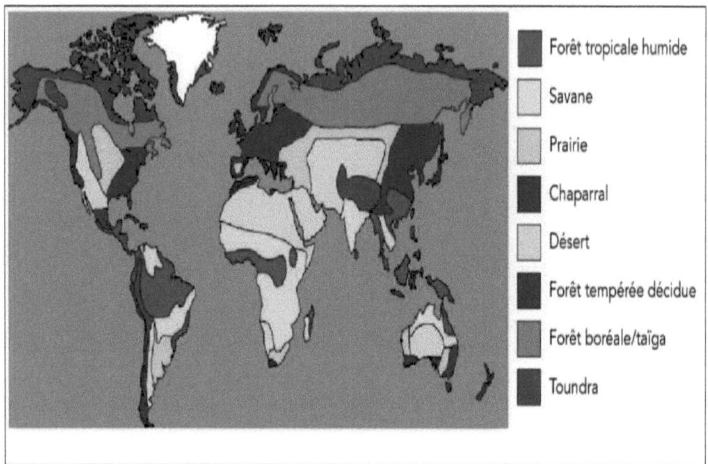

Figure 5: Distribution of the main terrestrial biomes.

Conclusion

The concrete delimitation of ecological systems depends on the objective of the study and the state of knowledge in the field. For example, we may be interested in the dynamics of a particular population, and so define the system to be studied by the network of direct or indirect relationships it has with the other biotic and abiotic components of its environment.

Chapter 2. The environment and its elements

Introduction

The influence of ecological conditions, in particular <u>habitat</u> characteristics, on the evolution of life histories is a central issue in evolutionary ecology. We can begin to address this question by considering the influence of <u>habitat</u> on the expression of trait trade-offs, with a bit of simple formalization of evolutionary trade-off optimization.

2.1. Notion of ecological niche

The concept of the ecological niche was coined by GRINELL in 1917 and later popularized by Elton in 1927. Organisms of a given species can maintain viable populations only under a certain range of conditions, for particular resources, in a given environment and during particular periods. The intersection of these factors describes **the niche**, which according to CAMPBELL & REECE (2007), is the position the organism occupies in its environment, including the conditions in which it is found, the resources it uses and the time it spends there (Figure, 6).

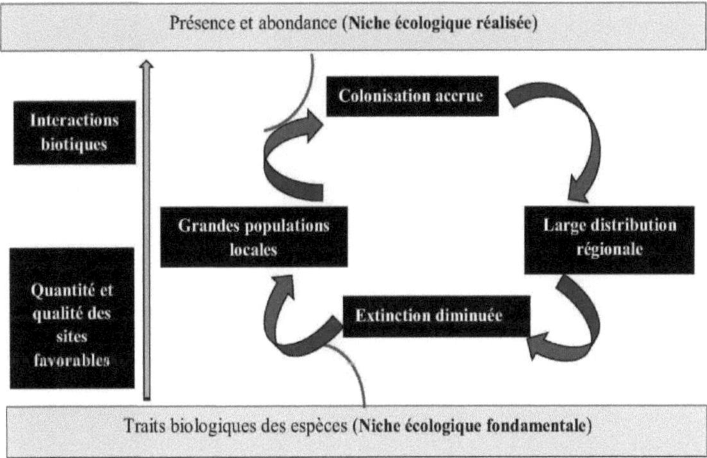

Figure 6. Community structuring through differences in ecological niches, with a particular role given to abiotic and biotic environmental filters and, through population dynamics, through colonization and extinction events (**after Verbek *et al.,* 2010**).

Organisms can change niches as they develop.

Example: Common Toads occupy an aquatic environment (feeding on algae and detritus) before metamorphosing into adults, where they become terrestrial (feeding on insects).

Table 1. Changes in the diet of toads and butterflies during their development.

	Example 1. Common toads	
Stadium	**Young**	**Adult**
Environment	Aquatic	Land
Power supply	Algae + garbage	Insects
	Example 2. Butterflies	
Stadium	**Larva**	Adult
Environment	**Foliage**	Flowers
Power supply	**Herbivore**	Nectarivore

Source: Ramade, 2009

2.2. *Notion of habitat*

Unlike a niche, an organism's habitat is the physical environment in which it is found.

Habitats contain many niches and support many different species.

Example: A forest has a vast number of niches for a choice of birds (nuthatches, woodcock), mammals (wood mice, foxes), insects (butterflies, beetles, aphids) and plants (wood anemones, mosses, lichen).

2.3. *Notion of environmental factors*

An "ecological factor" is any element of the environment that can have a direct effect on living beings.

Ecological factors are of two types:

Abiotic factors: of a physical or chemical nature, they have an enormous influence on various populations. We can cite climatic factors (temperature, humidity and rainfall, lighting and photoperiod, wind, etc.); edaphic, which concern the physico-chemical characteristics of soils (texture, structure, mineral elements present in the soil);

topographical, linked to relief; hydrological relating to aquatic biotopes (represented by the water's content of mineral salts, oxygen and other dissolved gases, by the current, the pH of the water, etc.) (RAMADE, 2009 & TIRARD *ETAL.,* 2012).

Biotic factors: we use this term to refer to all plant and animal populations, including microbes, whose action can maintain or modify ecosystem functioning.

We will first consider the plant population with floristic surveys, then the animal population with techniques for observing, capturing, counting, sorting and conserving fauna (FAURIE *ETAL.,* 2012).

2.4. *Interaction between the environment and living beings*

The reactions of living beings to variations in the physico-chemical factors of the environment concern morphology, physiology and behavior (RAMADE, 2008).

Living beings are either eliminated altogether, or their numbers are drastically reduced when the intensity of ecological factors is close to or exceeds tolerance limits.

2.4.1. *Tolerance law (tolerance interval)*

Enunciated by Shelford in 1911, the law of tolerance stipulates that for any environmental factor there exists a range of values (or tolerance interval) within which any ecological process dependent on this factor can operate normally. It is only within this interval that the life of a given organism, population or biocenosis is possible. According to TIRARD *ET AL.* (2012), the lower limit along this gradient delimits death by deficiency, while the upper limit delimits death by toxicity. Within the tolerance interval, there is an optimum value, known as the "preferendum" or "ecological optimum", at which the metabolism of the species or community in question proceeds at maximum speed (Figure 7).

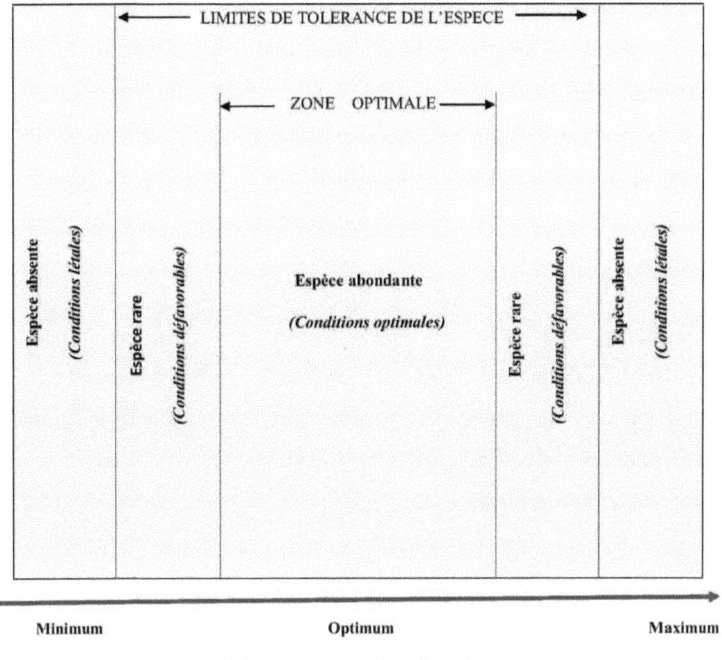

Figure 7. Tolerance limits of a species as a function of the intensity of the ecological factor studied. (Species abundance is highest in the vicinity of the ecological optimum).

A species' ecological valence represents its ability to withstand greater or lesser variations in an ecological factor. It represents the capacity to colonize or populate a given biotope (TRIPLET, 2015).

- A species with high ecological valence, i.e. capable of populating very different environments and withstanding major variations in the intensity of ecological factors, is said to be *Eurycecious.*
- A species with low ecological valence will only be able to withstand limited variations in ecological factors.
- A species with medium ecological value is called *Mesoèce.*

2.4.2. *Law of minimum*

Liebig (1840) coined the law of the minimum, which states that plant growth is only possible when all the elements required for it are present in sufficient quantities in the soil. It is the deficient elements (whose concentration is below a minimum value) that condition and limit growth (SOTTIAUX, 2008 & RAMADE, 2009).

Liebig's law is generalized to all ecological factors in the form of a law known as the "law of limiting factors".

2.4.3. *Limiting factor*

An ecological factor plays the role of a limiting factor when it is absent or reduced below a critical threshold, or when it exceeds the maximum tolerable level. According to TRIPLET (2015), it is the limiting factor that will prevent the installation and growth of an organism in an environment.

Conclusion

The concept of ecological niche is an often debated one, which goes beyond the notion of habitat, since it describes the way in which a species interacts with the whole of its biotic and abiotic environment. As a result, various ecological niches can be formed through a wide range of ecological interactions and on variable timescales, making them difficult to study.

Chapter 3. Abiotic factors

Introduction

Abiotic ecological factors represent constraints for organisms when their values are not within the optimum range for the species.

Despite severe environmental constraints, living organisms have successfully colonized the most hostile regions, be they poles, mountains, deserts or the deep sea. Their maintenance is made possible by responses at different levels: morphoanatomical, biochemical, physiological and behavioral.

3.1. Climatic factors

3.1.1. Defining climate

L limate is the set of atmospheric and meteorological conditions specific to a region of the globe. The climate of a region is determined from the study of meteorological parameters (temperature, humidity, precipitation, wind strength and direction, sunshine duration, etc.) evaluated over several decades (CHEMERY, 2004 & FAURIE *ET AL*, 2012).

3.1.2. Main climatic factors

The elements of climate that play an ecological role are numerous. RAMADE (2003, 2009), states that the main ones are temperature, humidity and rainfall, illuminance and photoperiod (the ratio of daylight to darkness during the day). Others, such as wind and snow, are less important, but in some cases can play a significant role.

3.1.2.1. Temperature

According to FRONTIER *ET AL.* (2008), temperature is the most important climatic element, as all metabolic processes depend on it. Phenomena such as photosynthesis, respiration and digestion follow Van't Hoff's law, which states that the speed of a reaction is a function of temperature.

The vast majority of living creatures can only survive in a temperature range between 0 and 50°C on average. Temperatures that are too low or too high trigger a state of dormancy (quiescence) in some animals, known as estivation or hibernation. In both cases, development is virtually halted.

The limits of geographic ranges are often determined by temperature, which acts as a limiting factor. Very often it is temperature extremes rather than averages that limit a species' ability to establish itself in an environment.

3.1.2.2. *Humidity and rainfall*

Water accounts for 70-90% of the tissues of many living species. Water supply and loss reduction are fundamental ecological and physiological issues. Depending on their water requirements, and consequently their distribution in the environment (BEAUX, 2004), we distinguish :

- Aquatic species that live permanently in water (e.g. fish);
- Hygrophilous species that live in damp environments (e.g. amphibians);
- Mesophilic species with moderate water requirements that can withstand alternating wet and dry seasons;
- Xerophilous species that live in dry environments where water deficit is accentuated (desert species).
- Living beings adapt to drought in a wide variety of ways:

In plants

- Reduced evapotranspiration through the development of impermeable cuticular structures.
- Reduction in the number of stomata.
- Reduced leaf surface, transformed into scales or spines.
- The leaves fall off in the dry season and reform after each rain.
- The plant obtains its water supply from a powerful underground apparatus.
- Water storage in aquifer tissues combined with good epidermal protection.

In animals

- Use of water contained in food.
- Reduced water excretion through increasingly concentrated urine.
- Use of metabolic water formed by fat oxidation (dromedary).

3.1.2.3. *Light and sunshine*

Following the observations of MACKENZI *ET AL.* (2000), we can say that sunshine is defined as the time during which the sun shines. Solar radiation is essentially composed of visible light, infrared rays and ultraviolet rays. Illuminance is important not only for its intensity and nature (wavelength), but also for the duration of its action (photoperiod). The photoperiod increases from the Equator towards the Poles.

At the Equator, days are strictly equal to nights throughout the year. At the Tropics, the inequality remains slight and practically without influence. At very high latitudes, i.e. beyond the Arctic Circle, nights and days exceed 24 hours, reaching 6 months of days and 6 months of nights at the Poles themselves. The atmosphere acts as a screen or, better still, a filter, blocking out some radiation while letting others through. In effect, the atmosphere absorbs part of the sun's rays, and diffuses another portion. To these two actions must be added the phenomenon of reflection.

Action on plants

Plants are adapted to the intensity and duration of light. This adaptation is important when plants move from the vegetative stage (growth and development) to the reproductive stage (flowering).

Plants can be divided into three categories:

- **Short-day plants:** they will only flower if the photoperiod at bud burst is less than or equal to 12 hours of light.
- **Long-day plants:** which need at least 12 hours of light to flower.
- **The indifferent:** light duration plays no role in flowering.

Action on animals

In animals, photoperiod plays an essential role in maintaining seasonal, daily (circadian) or lunar biological rhythms.

- **Seasonal biological rhythms:** there are two types:
    - **Reproductive rhythms in vertebrates:** the result is that the breeding period coincides with the favorable season.
    - **Diapause:** photoperiod is the key factor triggering the animal's entry into diapause

before the onset of the unfavorable season.

- **Daily or circadian rhythms**

These rhythms have a period of 24 hours. They are maintained by a poorly understood internal mechanism known as the "biological clock", whose regulation is conditioned by light and temperature.

- **Lunar rhythms**

These are activity rhythms triggered by moonlight. They are best known in marine animals.

3.1.2.4. Wind

Wind results from the movement of the atmosphere between high and low pressure (PARCEVAUX & HUBER, 2007). The impact of this factor on living beings can be summarized as follows:

- It has a drying effect because it increases evaporation.
- It also has considerable cooling power.
- Wind is a dispersal agent for animals and plants.
- Insect activity is slowed by the wind.
- When gales blow down trees in the forest, they create clearings in which young trees can grow.
- Wind has a mechanical effect on plants, which lie on the ground and take on particular shapes called anemomorphosis.

3.1.2.5. Snow

This is an important ecological factor in the mountains. Snow cover protects the ground from cooling. Under a metre of snow, the ground temperature is -0.6°C, compared with -33.7°C at the surface (RAMADE, 2008).

3.2. Edaphic factors

3.2.1. Soil definition

Soil is a complex and dynamic living environment, defined as the natural surface formation, with a loose structure and variable thickness, resulting from the transformation of the underlying parent rock under the influence of various processes: physical, chemical and biological, in contact with the atmosphere and living beings. It is made up of a mineral fraction and organic matter. Plants and animals draw water and mineral salts from the soil, and find the shelter and/or support they need to flourish.

3.2.2. *Edaphic factors*

3.2.2.1. *Soil texture*

Soil texture is defined by the size of the particles of which it is composed: gravel, sand, silt, clay (granulometry: measurement of the shape, size and distribution into different classes of grains and particles of divided matter):

Table 2. Soil texture as a function of particle size

Particle	Diameter
Gravel	>2 mm
Coarse sand	2 mm to 0.2 mm
Fine sands	0.2 mm to 20 μm
Loam	20 μm to 2μm
Clay	< 2μm

Source: Dadjoz, 2006

Depending on the proportion of these different particle size fractions, the following textures are determined:

- **Fine textures:** contain a high clay content (>20%) and correspond to so-called "heavy" soils, which are difficult to work, but offer optimum water retention.

- **Sandy or coarse textures:** these characterize light soils that lack cohesion and tend to dry out seasonally.

- **Medium textures:** there are two types:

  - Sandy-clay loam, which contains no more than 30 to 35% silt, has a perfectly balanced texture and corresponds to the best "free" soils.
  - Silt-textured soils, which contain more than 35% silt, are poor in humus (soil organic matter derived from the partial decomposition of animal and plant matter).

Biologically speaking, granulometry plays a role in the distribution of animals and groundwater. Many organisms, such as earthworms, prefer silty or sandy-clay soils, as do some species of beetle, which prefer clay and/or silty soils with a high content of fine elements, which are able to retain the necessary water, unlike coarse elements, which allow the soil to dry out too quickly.

3.2.2.2. *Soil structure*

Structure is the organization of soil. It is also defined as the spatial arrangement of sand, silt and clay particles (LACOSTE & SALANON, 2006). There are three main types of structure:

- **Particular:** where soil elements are not bound, the soil is very loose (sandy soils).
- **Massive:** where soil elements are bound together by cements (organic matter, limestone) to form a highly resistant discontinuous or continuous mass (clayey soils). This type of soil is compact and not very porous. It does, however, prevent temperature- and humidity-sensitive animals from migrating vertically, thus prohibiting their existence.
- **Fragmentary:** where elements are bound together by organic matter to form aggregates (heterogeneous assembly of substances or elements that adhere firmly to one another) of varying sizes. This structure is the most favorable for living beings, as it contains a sufficient proportion of voids or pores to support root life and biological activity in general, by allowing air and water to circulate.

3.2.2.3. *Soil water*

Water is present in the soil in four specific states:

- **Hygroscopic water:** comes from atmospheric humidity and forms a thin film around soil particles. It is retained very vigorously and cannot be used by living organisms.
- **Non-absorbable capillary water:** occupies pores smaller than 0.2 mm in diameter. It is also too energetically retained to be used by living organisms. Only certain highly adapted organisms can use it.
- **Absorbable capillary water:** located in pores measuring between 0.2 and 0.8mm. It is absorbed by plants and enables the activity of bacteria and small protozoa such as flagellates.
- **Gravity water:** temporarily occupies the largest pores in the soil. This water flows under the action of gravity.

3.2.2.4. *Soil pH*

Soil pH is the result of various soil factors. The soil solution contains H^+ ions from :

- Alteration of bedrock
- Humification of organic matter (synthesis of humic acid)
- Biological activity

- The effect of acidifying fertilizers

The pH also depends on the nature of the plant cover and climatic conditions (temperature and rainfall):

- basic pH (above 7.5) characterizes soils developed on calcareous parent rock. They are generally found in dry or seasonally dry climates and under vegetation with rapidly decomposing leaves.
- Acidic pH levels (between 4 and 6.5) are much more common in humid and cold climates, which are conducive to the accumulation of organic matter. They characterize coniferous forests. They form mainly on siliceous and granitic rocks (COUDURIER *ET AL.*, 2012).

Living organisms such as protozoa can withstand pH variations from 3.9 to 9.7, depending on the species: some are **acidophilic**, while others are **basophilic**. **Neutrophiles** are the most common in nature.

3.2.2.5. *Chemical composition*

Soil types vary widely in their chemical composition. The elements most closely studied in terms of their effect on flora and fauna are chlorides and calcium.
Saline soils, with their high sodium chloride content, have a very specific flora and fauna. Salt soil plants are **halophytes**.

Depending on their preferences, plants are classified as **calcicoles** (species able to tolerate high levels of limestone), and **calcifuges** (species able to tolerate only small traces of calcium) (RAMADE, 2009).

As for animals, calcium is necessary for many soil animals.
Abnormal soils contain high concentrations of more or less toxic elements: sulfur, magnesium, etc. Heavy metals exert a toxic action on vegetation. Heavy metals exert a toxic action on vegetation, leading to the selection of **toxico-resistant** or **metallophyte** species forming specific plant associations.

Conclusion

The responses of organisms to multiple environmental variations often take the form of adjustments implemented during acclimatization phases, rather than abrupt modifications. Acclimatization is the expression of phenotypic flexibility, which enables a gradual shift in the tolerance range of species, for example upwards or downwards with a progressive rise or fall in temperature respectively.

Chapter 4. Biotic factors

Introduction

Biotic factors are all the actions that living organisms exert directly on one another. These interactions, known as coactions, are of two types:

- **Homotypic** or intraspecific, when they occur between individuals of the same species.

- **Heterotypic** or interspecific, when they occur between individuals of different species.

4.1. Homotypic coactions

4.1.1. The group effect

A group effect is when changes occur in animals of the same species, when they are grouped in twos or more. The group effect is known in many insect or vertebrate species, which can only reproduce normally and survive when represented by large enough populations (TIRARD *ET AL*, 2016).

Example: It's estimated that a herd of African elephants needs to contain at least 25 individuals to survive: fighting enemies and finding food is made easier by living together.

4.1.2. The mass effect

In contrast to the group effect, the mass effect occurs when the environment, often overpopulated, provokes severe competition with harmful consequences for individuals (RAMADE, 2009).

The adverse effects of this competition have consequences for the metabolism and physiology of individuals, resulting in disturbances such as lower fertility rates, reduced birth rates and increased mortality. In some organisms, overcrowding leads to phenomena known as **self-elimination.**

4.1.3. Intraspecific competition

This type of competition can occur at very low population densities, and manifests itself in very different ways:

- Appears in territorial behavior, i.e. when the animal defends a certain area against incursions by other individuals.
- Maintaining a social hierarchy with dominant and dominated individuals.
- Food competition between individuals of the same species is intense when population density becomes high. The most frequent consequence is a drop in the population growth rate.

In plants, intraspecific competition, linked to high densities, is mainly for water and light. The result is a reduction in the number of seeds formed and/or high mortality, which greatly reduces the number of plants.

4.2 Heterotypic interactions

The cohabitation of two species can have a neutral, favorable or unfavorable influence on each of them.

4.2.1. Neutralism

Neutralism means that the two species are independent: they coexist without any influence on each other.

4.2.2. Interspecific competition

Interspecific competition can be defined as the active search by members of two or more species for the same environmental resource (food, shelter, egg-laying site, etc.) (CHAPIN *ETAL.*, 2012).

In interspecific competition, each species has an adverse effect on the other. The closer two species are to each other, the greater the competition.

However, two species with exactly the same needs cannot coexist, as one of them is bound to be eliminated after a certain period of time. This is the GAUSE principle, or the principle of competitive exclusion.

4.2.3. Predation

According to TIRARD *ET AL.* (2012), a predator is any free-living organism that feeds at the expense of another. It kills its prey in order to eat it. Predators can be polyphagous (preying on a large number of species), oligophagous (feeding on a few species), or monophagous (subsisting on a single species).

4.2.4. *Parasitism*

The parasite is an organism that does not lead a free life: at least at one stage of its development, it is bound to the surface (ectoparasite) or inside (endoparasite) of its host (CAMPBELL & REECE, 2007).

Parasitism can be considered a special case of predation. However, the parasite is not really a predator, as its aim is not to kill the host (TIRARD *ET AL.,* 2016). The parasite must adapt to meet the host and survive at the latter's expense. The host must adapt to avoid encountering the parasite and get rid of it if the encounter has taken place. Like predators, parasites can be polyphagous, oligophagous or monophagous.

4.2.5. *Commensalism*

Interaction between a commensal species, which benefits from the association, and a host species, which neither benefits nor harms. The two species interact through mutual tolerance.

Example: Animals that settle in and are tolerated in the roosts of other species.

4.2.6. *Mutualism*

It is an interaction in which both partners find a benefit, which may be protection against enemies, dispersal, pollination, nutrient supply, etc. (FRONTIER *ETAL.,* 2008).

Example: Tree seeds must be dispersed far and wide to survive and germinate. This dispersal is the work of birds, monkeys, etc., which benefit from the tree (food, shelter, etc.).

The obligatory and indispensable association between two species is a form of mutualism, to which we reserve the name symbiosis. In this association, each species can only survive, grow and develop in the presence of the other.

Example: Lichens are formed by the association of an alga and a fungus.

4.2.7. *Amensalism*

It's an interaction in which one species is eliminated by another that secretes a toxic substance. In plant interactions, amensalism is often referred to as **allelopathy**.

Example: The Walnut tree releases a volatile toxic substance through its roots, which explains the lack of vegetation under this tree.

Conclusion

Just as there are all kinds of interactions (competition, cooperation, predation, etc.) between individuals within populations, different species living side by side in the same community can have interactions with each other that can modify their dynamics or guide their evolution. All these intra- and interspecific relationships, changing in space and time, form complex networks within ecosystems.

Chapter 5. Ecosystem structure and function

Introduction

The study of ecosystems is one of the major objectives of ecology. It is by understanding the fundamental mechanisms of their functioning and equilibrium that rational bases for the conservation and management of natural heritage can be proposed. It is also through the integration of knowledge acquired on ecosystems and their interrelationships that the organization and evolution of the biosphere can be understood, reconstituted and controlled.

5.1. *The biosphere and its components*

Biosphere literally means sphere of life, i.e. all life on earth. Living beings are located in a narrow layer on the Earth's surface. This includes **the lower atmosphere,** the oceans, seas, lakes and rivers known as the **hydrosphere**, and the thin surface layer of land called the **lithosphere** (RAMADE, 2009 & TRIPLET, 2015).

The thickness of the biosphere varies considerably from one point to another, with life penetrating into the oceanic trenches beyond a depth of 10,000 m, whereas in the lithosphere, there is hardly a trace of life beyond a depth of around ten meters. In the atmosphere, due to the increasing scarcity of oxygen, living beings become rarer with altitude and rarely live above 10,000 m.

The major source of energy in the biosphere is the sun. The other major source is geothermal energy. Through photosynthesis, plants convert solar energy into chemical energy, which animals recover by eating the plants or each other.

5.2. *Biosphere organization*

The most elementary level of organization in living organisms is the cell. The cell is integrated into the individual, which in turn is integrated into a population. The population is part of a community or biocenosis. The biocenosis is in turn part of the ecosystem. Together, these ecosystems form the biosphere, the highest level of living organisms.

An ecosystem is made up of all living beings (biocenosis) and the environment in which they live (biotope).

The biotope provides energy and organic and inorganic matter of abiotic origin. The biocenosis comprises three categories of organisms: **producers** of organic matter, **consumers** of this matter and **decomposers** who recycle it. Plants capture solar energy and produce carbohydrates, which are then transformed into other products. They are grazed by **herbivores**, which are then eaten by **carnivores**. **Decomposers** consume the waste products and cadavers of all organisms, returning various substances to the environment. Through its unity, organization and functioning, the ecosystem appears to be the basic link in the biosphere (FAURIE *ETAL.,* 2012).

5.3. The trophic chain

5.3.1. Definitions

A trophic chain or food chain is a succession of organisms, each of which depends on the previous one. Every ecosystem comprises a set of animal and plant species that can be divided into three groups: **producers, consumers and decomposers** (DACHIN & GIRALDEAU, 2005).

5.3.1.1. The producers

These are photosynthetic autotrophic plants (green plants, phytoplankton: cyanobacteria or blue-green algae: prokaryotic organisms). According to GREULICH (2016), this trophic level forms the basis of the ecosystem's food chain. In fact, thanks to photosynthesis, they elaborate organic matter from strictly mineral matter supplied by the abiotic external environment.

5.3.1.2. Consumers

These are heterotrophic living beings that feed on the complex organic matter they extract from other living beings. They consider themselves to be secondary producers. Consumers occupy a different trophic level depending on their diet (SOTTIAUX, 2008 & GREULICH, 2016). We distinguish between consumers of fresh matter and consumers of corpses.

a- Consumers of fresh produce include :

- **Primary consumers (C1)**: These are the phytophagous that eat the producers. They are generally animals, called herbivores (herbivorous mammals, insects, crustaceans: shrimp), but also more rarely plant and animal parasites of green plants.

- **Secondary consumers (C2)** : Predators of C1. These are carnivores that feed on herbivores (carnivorous mammals, birds of prey, insects, etc.).
- **Tertiary consumers (C3)** : Predators of C2. These are carnivores that feed on carnivores (insectivorous birds, birds of prey, insects, etc.).

In most cases, consumers are omnivores and therefore belong to several trophic levels.

C_2s and C_3s are either predators that capture their prey, or animal parasites.

b- Consumers of animal carcasses

Scavengers are species that feed on the corpses of fresh or decomposed animals. They often complete the work of carnivores. **Example:** Jackal, Vulture, ...

5.3.1.3. *Decomposers or detritivores*

Decomposers are the various organisms and microorganisms that attack cadavers and excreta and gradually break them down, ensuring the gradual return to the mineral world of the elements contained in organic matter (SOTTIAUX, 2008 & FAURIE *ET AL,* 2012).

- **Saprophyte:** Plant organism feeding on decomposing organic matter.
  Example: Mushrooms.

- **Saprophage:** Animal that feeds on decomposing organic matter.
  Example: Bacteria.

- **Detritivore:** Invertebrate that feeds on animal and/or plant detritus or debris.

  Example: Protozoa, earthworms, nematodes, woodlice.

- **Coprophage:** Animal that feeds on excrement.
  Example: Dung beetle.

Primary producers, consumers and decomposers are linked by a food chain. The cyclical nature of the chain is ensured by the decomposers.

5.3.1.4. *Nitrogen fixers*

They have a special position in the trophic chain. Their nitrogen nutrition is based on molecular nitrogen. As for the carbon and energy required for their nutrition, they use more elaborate organic matter, which they take from certain detritus or from the roots or leaves of autotrophs. They are therefore autotrophic in terms of nitrogen and heterotrophic in terms of carbon (RAMADE, 2008). This is the case for Azotobacter in non-symbiotic fixation and Rhizobium in symbiotic fixation.

5.3.2. *Different types of trophic chains*

There are three main types of linear trophic chains (FISCHESSER & DUPUIS-TATE, 2007 ; SOTTIAUX, 2008 & RAMADE, 2009):

- **Chain of predators**

  In this chain, the number of individuals decreases from one trophic level to the next, but their size increases (Elton's rule, enunciated in 1921).

  Example: (100) Producers + (3) Herbivores + (1) Carnivore.

- **Parasite chain**

  On the contrary, the trend is from larger organisms to smaller but increasingly numerous ones (Elton's rule is not verified in this case).

  Example: (50) Grass + (2) Herbivorous mammals + (80) Fleas + (150) Leptomonas.

- **Chain of detritus feeders**

  Goes from dead organic matter to increasingly small (microscopic) and numerous organisms (Elton's rule is not verified in this case).

  Example: (1) Corpse + (80) Nematodes + (250) Bacteria.

5.3.3. *Graphical representation of trophic chains*

The structure of biocenoses is usually illustrated using ecological pyramids, which are superimposed horizontal rectangles of the same height, but with lengths proportional to the number of individuals, biomass or energy present in each trophic level. These are known as number, biomass or energy pyramids (Figure, 8).

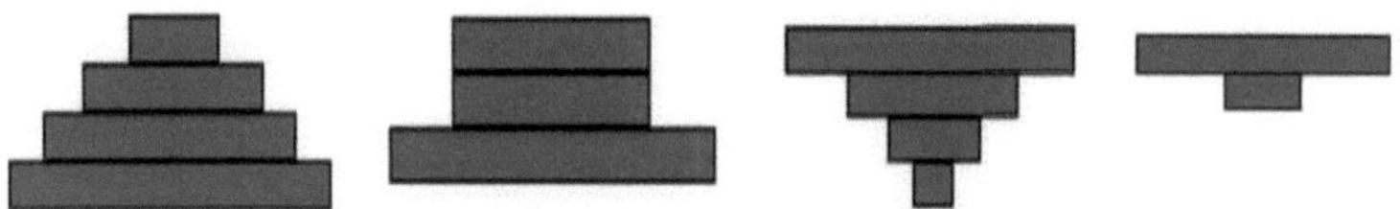

Figure 8. Various diagrams of ecological pyramids.

5.3.4. The food web

The food web is defined as a set of interconnected food chains within an ecosystem, through which energy and matter circulate. It is also defined as the set of trophic relationships existing within a biocenosis between the various ecological categories of living beings making up the biocenosis (producers, consumers and decomposers).

5.4. Energy transfer and efficiencies

5.4.1. Definitions

- **Gross productivity (GP):** Quantity of living matter produced per unit of time by a given trophic level.

- **Net productivity (NP) :** Gross productivity minus the amount of living matter degraded by respiration.

  PN = PB - R.

- **Primary productivity:** Net productivity of chlorophyll autotrophs.

- **Secondary productivity:** Net productivity of herbivores, carnivores and decomposers.

5.4.2. Energy transfer

The trophic relationships that exist between the levels of a food chain translate into energy transfers from one level to the next.

- Part of the sunlight absorbed by the plant is dissipated in the form of heat.
- The remainder is used for the synthesis of organic substances (photosynthesis) and corresponds to Gross Primary Productivity **(GP (BARBAULT, 2008 & TRIPLET, 2018).**
- Part of **(PB)** is lost to Breathing **(R1).**

- The remainder is Net Primary Productivity **(NP)**.

- Part of **(PN)** is used to increase plant biomass before falling prey to bacteria and other decomposers.

- The remainder of **(PN)** is used as food by the <u>herbivores</u>, which absorb a quantity of energy ingested **(I1)**.

- The quantity of energy ingested **(I1)** corresponds to what is actually used or Assimilated **(A1)** by the herbivore, plus what is rejected (Non Assimilated) **(NA1)** in the form of excrement and waste: **I1= A1+ NA1**

- The assimilated fraction **(A1)** is used on the one hand for Secondary Productivity **(PS1)** and on the other hand for Respiratory expenditure **(R2)**.

- The same reasoning applies to <u>carnivores</u>.

So, from the sun to the consumers (1er , 2ème or 3ème order), energy flows from trophic level to trophic level, decreasing with each transfer from one link to another. This is known as energy flow. The energy flow through a given trophic level corresponds to the total energy assimilated at that level, i.e. the sum of net productivity and substances lost through respiration.

In the case of primary producers, this flow is: **PB = PN + R1.**

The energy flow through the herbivore trophic level is: **A1 = PS1 + R2.**

The further away from the primary producer, the lower the production of living matter (Figure 9).

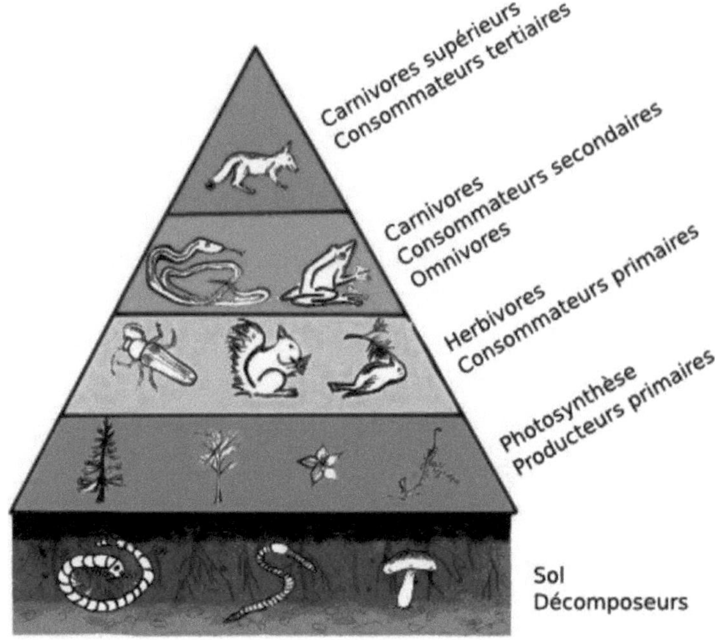

Figure 9. Biomass of the different levels of a food chain: moving from one food level to another entails a considerable loss of matter.

5.4.3. Yields

Energy is lost at every stage of the flow, from the eating organism to the eating organism, and within each of them. We can therefore characterize the various organisms from a bioenergetic point of view, by their ability to reduce these energy losses (FAURIE *ET AL.,* 2012). This ability is assessed by calculating yields:

- **Ecological yield:** This is the ratio of the net production of the trophic level of rank (n) to the net production of the trophic level of rank (n-1): **(PS1/PN x 100)** or **(PS2/PS1 x 100)**.

- **Exploitation efficiency:** Ratio of energy ingested **(I)** to energy available. It is the net production of the prey: **(I1/PN x 100)** or **(I2/PS1x 100).**

- **Net production yield:** This is the ratio of net production to assimilated energy: **(PS2/A2x100)** or **(PS1/A1x100).** This yield is of interest to breeders, as it expresses

40

the ability of a species to produce the greatest possible quantity of meat from a given quantity of feed.

5.4.4. *Ecosystem stability*

According to TRIPLET (2018), available resources, regulated by the physicochemical factors of the environment, control trophic chains from producers to predators. This is the theory of control of communities by resources (nutrients), or **bottom-up control.**

Example: The relationship between the phosphate content of the oceans + the quantity of plankton + the size of the fish that feed on them.

Conversely, the functioning of an ecosystem depends on the predation exerted by higher trophic levels on lower trophic levels. **This is top-down control.**

Example: Regulatory effect of a carnivore population (wolves) on a prey population (hares).

Both controls operate simultaneously in ecosystems and can be complementary. Human modification of a trophic level can amplify either of the two controls, leading to ecosystem instability.

Examples:

■ Increase in nutrient resources (amplification of bottom-up control). Organic water pollution or eutrophication.
Reduction in the abundance of a top predator (amplification of top-down control). Hunting and fishing.

5.4. *Biogeochemical cycles*

There is a circulation of matter in every ecosystem, with molecules or chemical elements constantly returning to their point of origin, which can be described as cyclical, unlike energy transfers. The alternating passage of elements, or molecules, between inorganic and living matter, is called the biogeochemical cycle. This cycle corresponds to a **biological cycle** (internal to the ecosystem, corresponding to exchanges between organisms), to which is grafted a **geochemical cycle** (a large-scale cycle, which may involve the entire biosphere and concerns transport in the non-living environment) (RAMADE 2008 ; RAVEN *ETAL.,* 2009 & TRIPLET, 2018).

There are three main types of biogeochemical cycles:

- The water cycle.
- The cycle of elements with a predominantly gaseous phase (carbon, oxygen, nitrogen).
- The cycle of elements with a predominantly sedimentary phase (phosphorus, potassium, etc.).

5.5.1. *The water cycle*

The water cycle involves the exchange of water between the Earth's various compartments: the hydrosphere, atmosphere and lithosphere (Figure 10).

Under the effect of the sun's heat, the water in seas, rivers and lakes evaporates. **Evapotranspiration** also plays an important role in the water cycle. It is accelerated by plants, which transpire large quantities of water through their leaf systems. What's more, their roots accelerate the upward movement of water in the soil-atmosphere direction. This water then joins the atmosphere in the form of water vapour (clouds).

Clouds are driven by the wind. When they cross cold regions, water vapor condenses (BEAUX, 2004; MUSY, 2005; PARCEVAUX & HUBER, 2007 & UNESCO, 2017). It falls back to earth in the form of rain, snow or hail. 7/9 of the total volume of this precipitation falls on the surface of the oceans, and only 2/9 on the continents. Water circulates in the lithosphere in three ways:

- **Runoff:** the phenomenon of water running off the surface of soils.
- **Infiltration:** the phenomenon of water penetrating into the ground through natural cracks in soil and rock, thus feeding the water table.
- **Percolation:** the migration of water through soils (to the water table).

Runoff, infiltration and percolation feed the watercourses that ultimately return the water to the hydrosphere.

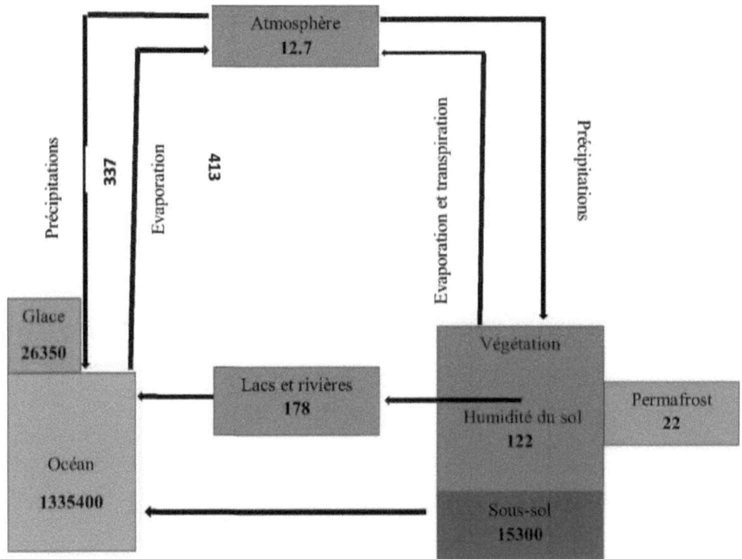

Figure 10. Global water cycle

(stocks are in thousands of Km3 and flows in thousands of Km3 per year).

5.5.2. *The carbon cycle*

During respiration, living beings consume oxygen and release carbon dioxide (CO_2) into the atmosphere. Similarly, industry and transport vehicles release CO_2 into the atmosphere after burning fuel in the presence of oxygen. Volcanic eruptions are also considered a natural source of CO_2. CO_2 is absorbed by plants (photosynthesis) and water (dissolution). Photosynthesis and dissolution are the phenomena that enable carbon dioxide to be recycled (Figure, 11).

After photosynthesis, carbon combines with other elements to form complex molecules, which after the plant's death are very slowly degraded into charcoal. When burned, these fossil fuels re-form CO_2.

The CO_2 in the air and that dissolved in water constitute the only source of inorganic carbon from which all the biochemical substances making up living cells are produced (thanks to chlorophyll assimilation).

During the respiration of autotrophs, heterotrophs and various other organisms, carbon dioxide is released in parallel with oxygen consumption.

CO_2 is also released during fermentations, which lead to partial decomposition of substrates under anaerobic conditions.

In soils, the carbon cycle often slows down: organic matter is not fully mineralized, but transformed into a set of acidic organic compounds (humic acids). In some cases, organic matter is not fully mineralized and accumulates in various sedimentary formations. As a result, the carbon cycle becomes stagnant or even blocked. This is currently the case with the formation of peat or, in the past, with the formation of large deposits of coal, oil and other fossil hydrocarbons.

But we're producing too much carbon dioxide, and the Earth can no longer recycle it. The level of CO_2 in the atmosphere is rising, and the climate is warming. This is because the CO_2 in the atmosphere traps the sun's heat, which makes life possible on Earth. This is known as the greenhouse effect. By increasing the concentration of CO_2 in the atmosphere, the balance of our ecosystem is disturbed. The climate warms up and this can have serious consequences for life on Earth: ice caps could melt and raise sea levels in certain places, causing flooding, increased extreme weather conditions such as storms, tidal waves, drought... etc. (CAMPBELL & REECE, 2007; BARBAULT, 2008; RAMADE, 2008; RAVEN *ET AL,* 2009 & UNESCO, 2017).

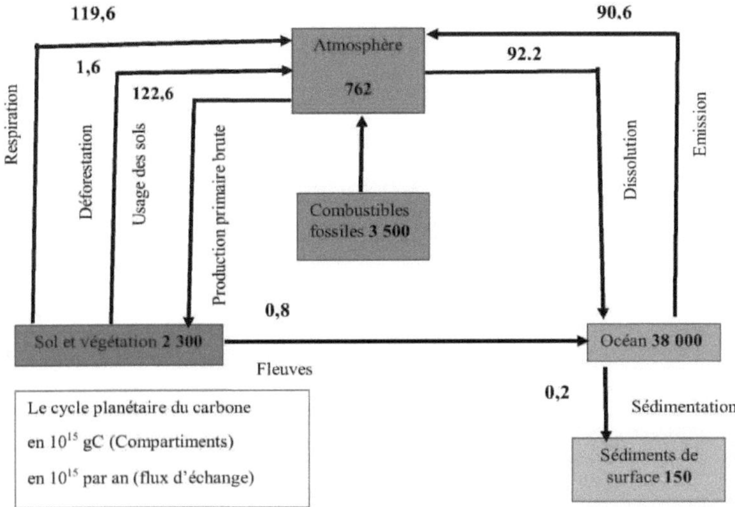

Figure 11. Global carbon cycle.

5.5.3. *The phosphorus cycle*

Despite the scarcity of mineral phosphorus in the biosphere, this element remains important for living matter (it is a constituent of DNA, RNA and ATP). Its main reservoir is made up of various rocks, which gradually give up their phosphates to ecosystems.

In the terrestrial environment, the concentration of assimilable phosphorus is often low and acts as a limiting factor. This phosphorus is circulated by leaching (or erosion) and dissolution, and thus introduced into terrestrial ecosystems, where it is absorbed by plants. Plants then incorporate it into various organic substances, passing it on to food webs. Organic phosphates are then returned to the soil with the corpses, waste and excreta produced by living beings, attacked by micro-organisms and transformed back into mineral orthophosphates, once again available to green plants and other autotrophs.

Phosphorus is introduced into aquatic ecosystems by runoff. This water then flows into the oceans, enabling the development of phytoplankton and animals in the various links of the trophic chain.
Phosphorus is converted from an organic to an inorganic state by bacteria and fungi.

A partial return of phosphates from the oceans to land occurs via **fish-eating** or **piscivorous** seabirds via guano deposits.

However, in the oceans, the phosphorus cycle takes place with losses, since a significant proportion of the phosphates carried out to sea are immobilized in deep sediments (fragments of fish corpses, not consumed by detritivores and decomposers). In the absence of upwelling currents, phosphorus scarcity is a limiting factor. The phosphorus cycle is therefore incomplete and open. Because of its scarcity and because of these losses to the cycle, phosphorus is therefore the main limiting factor controlling most primary production (CAMPBELL & REECE, 2007; RAMADE, 2008 & TRIPLET, 2018).

5.5.3. *The nitrogen cycle*

The main reservoir of nitrogen is the atmosphere, which contains 79% by weight. Inorganic nitrate formation is constantly taking place in the atmosphere as a result of electrical discharges during thunderstorms. But it plays only a secondary role to that of nitrifying micro-organisms. The latter are mainly represented by bacteria, either free-living (Azotobacter, Clostridium, Rhodospirillum) or symbiotic (Rhizobium). In the aquatic environment, cyanophyceae (blue-green algae) are the main fixers of nitrogen gas (Figure 12).

The nitrate nitrogen thus produced by these numerous terrestrial or aquatic micro-organisms is finally absorbed by plants, taken into the leaves and transformed into ammonia, thanks to a specific enzyme, nitrate reductase. The ammonia is then transformed into amino nitrogen and proteins.

The proteins and other forms of organic nitrogen contained in cadavers, excreta and organic waste are attacked by bioreducing microorganisms (bacteria and fungi) that produce the energy they need by decomposing this organic nitrogen, which is then transformed into ammonia - ammonification.

Some of this ammoniacal nitrogen can be absorbed directly by plants, but it can also be used by nitrifying bacteria (Nitrosomonas) to produce metabolic energy. These transform NH_4^+ ammonia into nitrite, NO_2^-, i.e. nitritation, and then Nitrobacter transform it into N03⁻ , i.e. nitratation. The nitrate ion N03⁻ is then absorbed by plants.

Nitrogen constantly returns to the air through the action of denitrifying bacteria (Pseudomonas), which are capable of breaking down the N03 ion⁻ into N2, which volatilizes and returns to the air; fortunately, however, the role of these bacteria is not very important.

A non-negligible proportion of nitrates can be leached by runoff and carried out to sea. Nitrogen can then be immobilized by incorporation into deep sediments. However, it is largely taken up by phytoplankton organisms and enters a food chain culminating in birds, which bring it back, through their droppings, to the terrestrial environment in the form of guano (RAMADE, 2009; FAURIE *ET AL,* 2012; TIRARD *ET AL,* 2016 & TRIPLET, 2018).

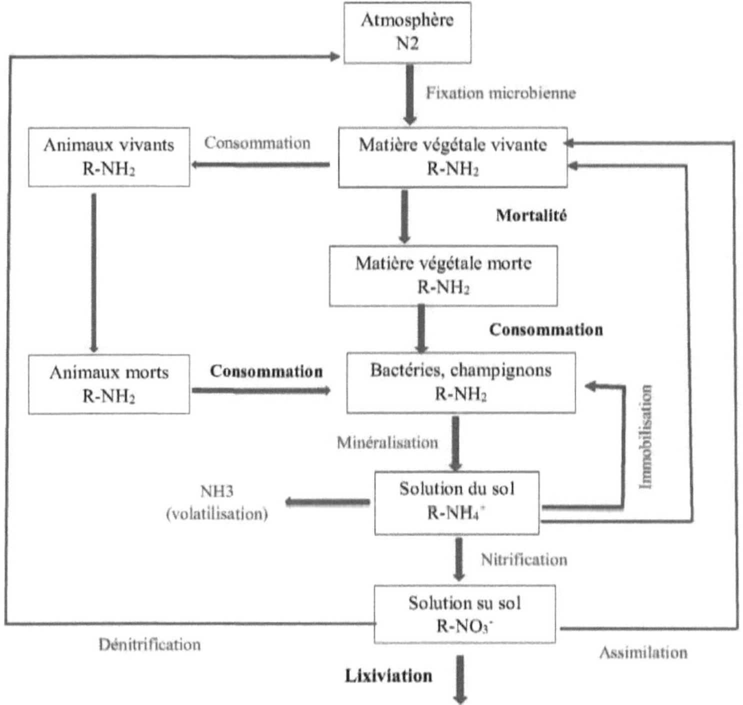

Figure 12. The nitrogen cycle in terrestrial ecosystems.

5.6. Impacts of human activities on ecosystem equilibrium

The impact of human activity on natural environments often has negative repercussions on various types of ecosystem. These can be subject to ecological imbalances of varying severity, depending on the scale of anthropogenic activity. Today, the vast majority of ecosystems, and therefore the entire biosphere, are subject to unprecedented anthropogenic pressure. Impressive technological advances have brought many benefits to human life, but in return they have had adverse repercussions on nature, such as pollution, reduced biodiversity, and the dysfunction and disappearance of many terrestrial and aquatic ecosystems.

5.6.1. Pollution of aquatic ecosystems

Water pollution is undeniably one of the most worrying aspects of the global environmental crisis. This pollution is caused by urban and industrial outfalls, but also by diffuse sources of contamination scattered over vast territories. The water crisis has been going on for a long time, and is getting worse, affecting both industrialized and Third World countries.

Responsibility for this pollution is attributed to urban and industrial emissions, as well as to diffuse sources of contamination dispersed over vast territories (RAMADE, 2005; KHALED-KHODJA, 2016 & UNESCO, 2017).

5.6.2. Air pollution

Gases, whether liquid or solid, in the atmosphere are present at levels high enough to cause harm to humans and other organisms or materials. Although atmospheric pollutants are sometimes produced by natural sources, such as lightning triggering a forest fire or the eruption of a volcano, air pollution is mainly caused by human activities that release all kinds of substances into the atmosphere.

It's clear that human-induced environmental degradation is mainly caused by the untimely release of various substances into the atmosphere. Although it dates back to the days of industrial civilization, air pollution has increased significantly in all developed countries.

As a result of increased industrial production and motor vehicle traffic, the air is increasingly polluted by fumes, toxic gases and other pollutants.

It would be futile to ascribe this rise in air pollution to any specific type of industrial activity or modern technology. In fact, it stems from a number of factors that define modern civilization: increasing energy production, the metallurgical industry, road and air traffic, incinerated waste tonnages - all of which are significantly involved in this phenomenon. (DELMAS *ETAL,* 2007 & RAVEN *ETAL.,* 2009).

5.6.2.1. The hole in the ozone layer

Ozone (O3) is an oxygen compound that is polluted by human activities in the troposphere, but is also an essential natural element in the stratosphere that surrounds our planet, at altitudes of between 17 and 50 km (RAVEN *ETAL.,* 2009).

Although ozone is present everywhere in the air, it is in the stratosphere that it is found in high concentrations, forming the ozone "layer" or screen. The latter stops most ultra-violet radiation, especially the shortest wavelength band (the most harmful). The protective function of this screen has enabled living beings to colonize the continents that have emerged (BOVET *ET AL,* 2008 & RAMADE, 2012).

Further chemical reactions release chlorine, which then metamorphoses into other ozone molecules in an endless catalytic cycle (CAMPBELL & REECE, 2007). In addition, the destruction of the ozone layer could have serious consequences for life on earth: increased cataracts, skin cancer and a weakened immune system. Ecosystems could be disrupted if UV levels rise.

5.6.2.2. Acid rain

Atmospheric pollution is having a dramatic impact on a continental scale, particularly due to the spread of acid rain, which is currently affecting almost the entire boreal hemisphere. The consequences for continental, aquatic and forest ecosystems are enormous (BEAUX, 2004). Acid rain occurs when the water content of precipitation is abnormally high compared with that from unpolluted environments. The acidity of this rain is essentially due to two atmospheric pollutants: 70% sulfur dioxide (SO2), abundantly produced by coal-fired power stations, the metallurgical and pulp and paper industries, and 30% nitrogen oxides (NOx), mainly from combustion and exhaust gases.

These gases mix with atmospheric humidity to produce sulfuric acid and nitric acid respectively. Organic acids can also be produced during hydrocarbon oxidation reactions.

Rain, snow or condensation (dew) then deposits these acids on the earth, whose pH is below the natural pH of 5.6 (BEAUX, 2004; CAMPBELL & REECE, 2007; DELMAS *ET AL.,* 2007). Acid rain causes considerable ecological damage. It is responsible for the acidification of many lakes and rivers. Acidification of water bodies leads to a significant decline in aquatic life, a reduction in the complexity of food webs and can, in the long term, cause the ecological death of the lake.

Another worrying aspect of acid rain is the significant deterioration of boreal and temperate forests. Conifers are the first to be affected by lesions, as their needles, which remain green all year round, turn yellow and fall off. The tree, whether coniferous or deciduous, gradually loses its leaves, dries out and dies on its feet (BEAUX, 2004 & RAVEN *ETAL.,* 2009). Acidified soils lead to a loss of nutrients (Ca, Mg, K, etc.) and a slowdown in the activity of decomposers (microfauna and microflora), resulting in a mineral shortage.

This will inhibit plant growth. Furthermore, an acidic environment favors the concentration and dissolution of heavy metals such as aluminum, which can lead to root poisoning (BEAUX, 2004 ; DELMAS *ETAL.,* 2007 & RAVEN *ETAL.,* 2009).

5.6.2.3. *Greenhouse effect*

The greenhouse effect is a natural mechanism that warms the atmosphere by absorbing the sun's rays through atmospheric gases. In the stratosphere, ozone is responsible for absorbing a fraction of ultraviolet radiation, while part of the visible (incident) light is reflected back into space by the atmosphere or the earth's surface. The energy lost or reflected is known as albedo.

Around half of the solar energy absorbed by the earth's surface is absorbed. Through radiation, the earth must lose energy to maintain thermal equilibrium. As a result, it radiates the energy received, but in a different wavelength range (invisible thermal infrared) from that of sunlight, which is mainly present in the visible range.

Little sunlight is absorbed by the atmosphere, while most of the infrared radiation re-emitted by the earth is absorbed by the atmosphere. This trapping of thermal radiation by the atmosphere is the earth's natural greenhouse effect. It is thanks to this effect that the earth's average surface temperature is 15°C; without it, it would be -18°C (BEAUX, 2004; CHEMERY, 2004; VALLEE, 2004; DELMAS *ET AL.,* 2007 & RAMADE, 2012).

Conclusion

Living organisms consume resources whose chemical nature is often very different from their own. In particular, the relative contents of the elements **water, carbon, nitrogen, oxygen, phosphorus**, etc. vary throughout the trophic chain, including at the level of decomposers of dead matter. These variations in elemental composition between organisms and between trophic levels create constraints that call for physiological, behavioral and demographic evolutionary responses.

General conclusion

One of the historic tasks of ecology is to analyze the causes, conditions and mechanisms by which the living world diversifies, as well as the interactions of this change with the physico-chemical environment.

This field of research is booming today, in the wake of the rapid disappearance of a large number of species. The questions are vast: is Biodiversity important? Does it play a role in regulating our environment and its capacity to produce renewable resources? Is there a biodiversity threshold below which we must not go, on pain of ecosystem degradation?

The question of community biodiversity - the diversity of species and their connections, the diversity of food webs - is of paramount importance today, not only for sociological and ethical reasons, but also because of the direct link between biodiversity and certain functional aspects of ecosystems, such as productivity and stability.

Determining areas of maximum diversity, and understanding the mechanisms that determine diversity levels, can help predict and manage future biodiversity.

"Ecology is also, and above all, a cultural problem. Respect for the environment requires a large number of behavioral changes.

Nicolas Hulot

References

- **BARBAULT R. 2008**. General Ecology. Structure et fonctionnement de la Biosphère. 6eme édition DUNOD.

- **BEAUX J.-F. 2004**. L'environnement. Nathan Éd., France. 160 p. - Beaux, 2004;

- **BOVET P., REKACEWICZ P., SINAÏ A. & VIDAL D. 2008**. L'Atlas de l'environnement.
Éd. Armond Colin, Paris. 103 p.

- **CAMPBELL N. & REECE J. 2007**. Biology. 7th ed. Pearson Éducation France, 1334 p.

- **CHAPIN III F.S., MATSON P. & VITOUSEK P.M. 2012.** Principles of terrestrial ecosystem ecology. 2nd edition. *Springer.*

- **CHEMERY L. 2004**. Petit atlas des climats. Petite Encyclopédie, Larousse. 128 p. - Chémery, 2004;

- **COUDURIER C., BOURGOGNE A. & TOUSSAINT H. 2012**. Guide pédologique, les sols. Alterre Bourgogne, France. 32 p.

- **DACHIN E., GIRALDEAU L.A. & CEZILLY F. 2005.** Behavioral ecology. Cours et question de reflexion. Dunod.

- **DAJOZ R. 2006**. Précis d'Ecologie. 8ème edition. Edition Dunod. 631p.

- **DE PARCEVAUX S. & HUBERT L. 2007**. Bioclimatology: Concepts and applications. Collection Synthèses (INRA), Editions Quae, 324 p.

- **DELMAS R., CHAUZY S., VERSTRAETE J-M. & FERRE H. 2007**. Atmosphere, Ocean and Climate. Éd. Belin, Paris. 287 p.

- **EL ABOUDI A. 2014**. General ecology "Plant ecology". Université Mohamed V-Agdal, Faculté des Sciences, Rabat. 124 p.

- **FAURIE C., FERRA C., MEDORI P., DEVAUX J. & HEMPTINNE J.L. 2011.** : Ecologie : Approche scientifique et pratique. 5ème edition. Edition Lavoisier. 407p.

- **FISCHESSER B. & DUPUIS-TATE M-E. 2007**. The illustrated guide to ecology. QUAE Éd., France. 350 p.

- **FRONTIER S., PICHOD-VIALE D., LEPRETRE A., DAVOULT D. & LUCZAK C. 2008**. Ecosystèmes, Structure, Fonctionnement, Évolution. 4ème Éd. Dunod, Paris. 576 p.

- **GREULICH S. 2016**. Ecology - upgrading. École polytechnique de l'Université de

Tours, Département d'Aménagement et Environnement. UMR CNRS (IPAPE). 102p.

- **KHALED-KHODJA S. 20i6**. Evaluation of the physico-chemical quality of anthropogenic discharges (urban, agricultural and industrial) to the Gulf of Annaba. Doctoral thesis, University of Annaba. 186 p.

- **LACOSTE A. & SALANON R. 2006**. Elements of biogeography and ecology. 2nd ed. Armand Colin, Paris. 318 p.

- **MACKENZI A., BALL A. S. & VIRDEE S. R. 2000**. Essentials of Ecology. Berti Éd. Paris, 368 p.

- **MUSY A. 2005**. General hydrology course, chapter 1 summary. École Polytechnique Fédérale de Lausanne, France. 4 p.

- **RAMADE F. 2003**. Éléments d'Écologie, Écologie fondamentale. 3rd ed. Dunod, Paris. 690 p.

- **RAMADE F. 2005**. Éléments d'Écologie, Écologie appliquée. 6th ed. Dunod, Paris. 864 p.

- **RAMADE F. 2008.** Encyclopedic dictionary of natural sciences and biodiversity. Dunod Éd. Paris, 726 p.

- **RAMADE F. 2009**. Éléments d'Écologie, Écologie fondamentale. 4th ed. Dunod, Paris. 689 p.

- **RAMADE F. 2012**. Éléments d'Écologie, Écologie appliquée : action de l'homme sur la biosphère. 7th ed., Paris. 791 p.

- **RAVEN P.H., BERG L. R. & HASSENZAHL D. M. 2009**. Environment. De Boeck University, Brussels. 687 p.

- **SOTTIAUX B. 2008.** General and applied ecology course, course notes. Industrial and commercial courses. Couillet, Belgium. 31 p.

- **TIRARD C., ABBADIE L., LALOI D. & KOUBBI PH. 2016**. Ecologie, Licence, Master & CAPES. DUNOD, 11, rue Paul Bert, 92240 Malakoff. www.dunod.com.

- **TIRARD C., BARBAULT R., ABBADIE L. & LŒUILLE N. 2012.** Mini manual d'écologie. Dunod Éd. Paris, 157 p. - Tirard et al. 2016

- **TRIPLET P. 2015.** Dictionary of biological diversity and nature conservation. Baie de Somme, Grand Littoral Picard. France, 722 p.

-   **TRIPLET P. 2018**. Encyclopedic dictionary of biological diversity and nature conservation. 4th ed. Baie de Somme, Grand Littoral Picard. France, 1096 p.

-   **UNESCO, 2017**. United Nations Educational, Scientific and Cultural Organization. 2017, Educational kit on biodiversity. Vol. 1. UNESCO, Paris. 192 p

- **VALLEE J-L. 2004**. Techniguide de la Météo. Nathan Éd., Paris. 221 p.

Printed by Books on Demand GmbH, Norderstedt / Germany